# Inside A Human Mind

## Between science and beliefs

Authored By

Nooh Turk

<u>**To my beautiful Zaina**</u>

2

<u>**To my beautiful Zaina**</u>

وَ يَسْأَلُونَكَ عَنِ الرُّوحِ قُلِ الرُّوحُ مِنْ أَمْرِ رَبِّي وَ مَا أُوتِيتُم مِّن الْعِلْمِ إِلاَّ قَلِيلاً

Al-Isra Verse No:85

:And they ask thee about the revelation. Say: The revelation is by the commandment of my Lord, and of knowledge you are given but a little.

# Table of contents

# Prologue

Have you ever felt that your life is on a different page than others?

I do. But I am probably not the only one on earth who sees things differently, as I am not different than anyone else.

But I have been gathering some facts to put it in front of you, though I am not forcing you to believe in what I believe.

My name is Nooh Mohammed, and I am from Earth, a societies that believes the unbelievable, and denies the facts, the facts that are always in front of us, facts that make sense to us, but still many of my fellow humans choose to lie and believe their lies.

Those lies often lead to another lie, and lies lead to wars, and wars lead to losing people that we love, and losing people who we love leads to hate, hate will lead us to revenge, revenge will lead us to savagery, savagery will lead us to animalism, and animalism will lead us to… atheism.

In this book I will be talking about how the human mind works or thinks and some facts that human often reject in believing.

# Dawn Of Humanity

What is mind? A definition of thoughts, memory, intelligence, feelings and also a personality. These are the major function of our human mind. Although there is a brain in all animals, what makes us different is that human mind has the ability to think and choose.

No one came into this life knowing how to speak, walk, or eat. When you are first born, your brain is 100% ready to absorb information and actions to learn; the more you feed your brain with education the more it develops.

Our modern technology came from the ability of our brain to develop.

When I was at school. I always use to ask myself, why do we need math, we only need to know how to add and multiply, that's it, why are these teachers trying to give us useless information that we will never use in our daily lives.

I was mistaken, when I went to college, I knew how important was to find $X2$ or how to find the solution of "x − 4 = 10."

Although this equation might not help you in your daily life, it will help your brain to think and this process will help your brain develop and improve its cells. Consider that there are some brain cells that are not working anymore, or they are asleep, and this equation is an exercise to wake up your sleeping brain cells.

Now when you solve this equation, it might not be easy for you, but at the end, you will be reborn with stronger brain cells, which will help you think wiser and better in your daily life.

I wish they could of told us more about math when I was younger, I guess I was just a teenager at that time, and all I cared for was my happiness.

Feed your mind with the correct information, accept the facts that are around you even if it hurts you.

A friend of mind recently got a divorce, and he is going through a lot. I really wish I can help him, but he won't listen to me. He believes that his marriage was destroyed by people's curses, or that people envied his marriage.

I wish you could listen to me, my dear friend, if you are reading this. Your marriage was destroyed based on your selfish actions. You always wanted more, and you gave less. No one tried to destroy your marriage, you bought it on yourself.

Sometimes we deny accepting the facts because it hurts us too bad. That is why we blame our failures on things that might not exist.  If you got a D– on your English exam, don't blame it on your teacher, try to talk to yourself and see the cause of your weakness instead of blaming it on your friends or teachers.

Our lives can be brighter if every one of us does his job correctly. We could live in a world of peace with no hunger and war.

We were born to be wise, we were born to develop this earth to a better place. If you are looking for a change in this world, start it with yourself, then your family and friends.

Be the change that you want to see in the world.

Don't throw your mistakes on other people. For example, if you don't pray, then don't try to convince people to pray, fix yourself first, take full responsibility that you are where you are because you have chosen where you wanted to be.

Don't tell me where I have to be when you have no idea where you are.

Find your path first, build your foundations on strong real facts, then come and face me. It is a sad thing that we live in a world that all of us are wearing a mask called "hope" … hoping that tomorrow will be a better day. You will always be nothing as long as your mind is surrounded with Hope.

Don't just hope. Start working to make it a reality, and if it works someday, only then try to fix other people's lives.

Life is not about sitting behind your laptop or tweeting quotes trying to impress others, while you are nothing but a failure. Life is about improving yourself first, then improving the people surrounding you.

All these people that I know in my life who are always blaming others for their failure, they don't own anything in their life, all what they do is cause a problem and blame others. I deal with these people almost in my daily life.

# Beliefs About The Earth

From the beginning, humans used to hunt and fight for their livelihood. They did not understand our planet Earth; all what they knew was hunting and finding shelter.

When they had to leave to another location to make a better living, their knowledge about their small world was expanded, but still the place that they lived in was considered the entire universe to them.

When civilization started to grow and move outwards from Africa, people started trading and doing businesses together. Some of the goods they traded did not exist in their nation, so they had to expand their businesses to other lands. Soon they started visiting other civilizations and that led to exploration and seeing new cultures and becoming acquainted with different belief system.

But unfortunately sometimes it led to anger because they started to feel sensitive that there is another belief system that might be stronger than theirs.

One of the different beliefs between the nations of the past they had, is the shape of our planet Earth, and their images about planet Earth that they created in their mind was nothing but a superstition.

Those civilizations had scientists and people whose minds were open. Their minds helped them become educated and do calculations to discover more about our planet Earth.

Soon they understood that the earth is nothing but a huge ball swimming in space, and that it is only one planet among many in our solar system.

This exactly shows how beliefs can change and evolve over time. They used to look at the earth as if it was their entire and only universe. But today we look at the earth more like a grain of sand floating in endless space… that's because time has changed our mind and thinking. We know more about space today and we are able to combine our Beliefs with the Truth, and the result is knowledge.

In the past, people use to look at the space and study the planets and stars on the top of their temples. Most of them were priests and men of religion of that nations.

- The Babylonians

They use to believe that the earth is an empty mountain from the inside, and that mountain is carried on a sea, and that inside the empty mountain is the dark world of dead.

- The Egyptians:

The Egyptians believed that the earth is a women laying on her left shoulders, and that she is covered except for her head with grass and trees. And the sun that they considered it as a God, was her ship that she travels from the living world to the world of death into the space.

- The Indians:

The Indians imagines the planet earth like a pyramid, and the pyramid was built on an elephant, if it moved or walked, an earth quick starts to happen.

All of these civilizations had their own beliefs about the shape of our planet. Since the beginning of life on earth, human have had beliefs about everything, still they tried to live in peace between their nations.

Nowadays, most wars are not between nations; they are between the citizens within a nation. Most of these wars are caused by having different beliefs from one another.

---

# Beliefs vs. Facts

In the past people, used to believe that ghosts had several heads, and this is how they pictured Ghosts in their mind.

The actual meaning for believing that they have several heads is that ghosts are able to take shapes. The people of those earlier times were limited thinkers, so they chose to believe without understanding.

To believe in something should mean that you also understand the facts that are behind the belief. For example, you should not only believe what you have been told, you have to understand why you have been told that and what the facts are behind the belief.

The problem is, history is very important to all of us, but history can be changed, and manipulated. But to understand history we need to understand it carefully, and we must look through it carefully and read more than one history book.

History can be changed, for example, if you were told that the army of the Roman Empire had nine million men in it, then you should read books about the Roman Empire, and how large their population was at that time to see if it possible that the army contained nine million men.

This is the problem with our world that people come to believe anything they hear or read, without studying it and checking the facts.

# The Evolution of the Human Mind

I've always wanted to write a book, a simple one that can send my idea into the world.

I wanted the whole world to feel what I feel and think what I think. Even if they might disagree with me, it would be enough for me to know that I shared my thoughts with others.

Our mind consist of two parts, believer mind, and the unbeliever mind. Over the past 100 years we have been exploring and inventing technology that can make the world a better place, all of the technology you see now is the result of the hard work of the believer mind.

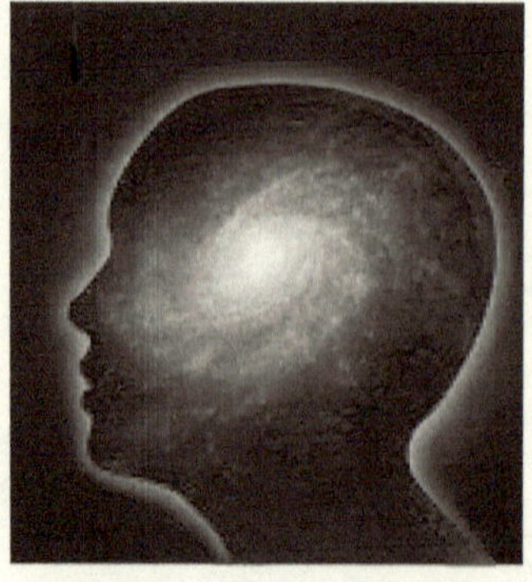

Take a look to our world now; we had nothing 100 years ago, all of these happened within one century, the question is, what happen to human mind?

The human mind for thousands of years was having a "NO" virus, which kept them from developing. Its only goal was to live and die.

We are the new generation, and we are the believers ... we can make the world a better place.  With science, life can be easier than it ever was, one of the best examples of technology is the importance of electricity, and science generates knowledge by means of new discoveries that are often met with beliefs.

Every school child knows that the earth revolves around the sun, this knowledge was rejected when it was first discovered, because people of that time did not want to admit that their beliefs were based on lies.

Look how close we made it, we fly from one continent to another within hours, which 1000 years ago took months to travel so far. Don't you think it's creepy? We are developing very fast!

The answer is... Evolution, our unbeliever father's mind was unable to update with the new data, but years after years the "NO" virus updated itself to a "YES" virus, and then they became believer minds, and accepted the fact that the human mind is able to develop and expand.

They started dreaming, achieving and hoping, they had faith and courage, that's how our mind evolved.

Evolution is everything in our life, without it nothing can make sense, you can never expect someone to be born with knowledge, but you can teach him day by day and feed him knowledge... "You were born to learn and write".

Try to learn and observe the actions of this life, even if it has nothing to do with your life style. The moment you see something that's unusual in your life, then try to have a moment and learn from it, you don't know, maybe one day you will be in the same situation, so feed your brain with the vaccination for the NO virus.

In the future you won't need to be worried about these situation, because once you vaccinate your brain, you can lay back and leave the rest for the mind to deal with it, and your auto memory will deal with it and give you easy answers.

But unfortunately people's mind is very different now days, no matter how much we develop, still some people choose to be unbeliever mind.

# Moon Landing Doubters

I live in a society where many people believe that the USA never made it to the moon.

They believe it's impossible to do that, though they never asked themselves if it is possible to travel from Bahrain to Dubai within 45 minutes.

If you ask them why they don't believe that USA made it on the moon, they will simply say that the American flag was waving but there is no air on the moon!?

What makes the flag flap if there's no wind on the moon?

You might have also see a few videos of the flag on the Moon waving back and forth. It actually happened when the astronauts first planted the flag… the flag was flapping.

So how did it happened?

There's no wind to make the flag wave, but there's also no wind to stop it from moving back and forth. When the astronauts planted the flag on the Moon, they couldn't help but to give it a sideways push.

Without the wind resistance the flag would experience on the Earth, but on the moon is a different case.

After the push the flag can flap back and forth a few times before finally settling down. That's why it looks like it's flapping, even though there's no wind.

So if NASA really landed on the moon, why did they stop going there?

I always get this question from people. The answer is simple. Space travel requires a lot of preparation and equipment.

Apollo program has a total cost of about $30 billion in 1970s which today would be more than $100 billion. So why did they stop going to the Moon? It's simple: because they would spend all that huge money on very little of interest on the Moon!  They can now spend far less sending robots, not humans, to the moon.

# The Pyramids—Proof of Human Intelligence, not Miracles

If you want to take a look at the pyramids, people say it is a miracle and how could they build such a huge pyramids like this, but I believe that this is just an art, an art that they were good at it, they were smart and intelligent nation, and also they were educated, I am sure they did the math correctly that time to build such big pyramids that have lasted for thousands of years. But as I said, it's not a miracle.

I believe they had knowledge and not super human power or magic as some people believe.

But still it's funny when I remember that some of the people call the pyramids a Miracle, knowing that they had no technology in their time, but still they don't believe that we landed on the moon with all the technology surrounding us.

Imagine if you bring someone from the past, show him what we have, all these high buildings we have, bridges and airplanes, what would you aspect him to say?

He would say that it's a magic or a miracle, it will take you a life time just to persuade him that all of these building and airplanes was built with knowledge and not magic as he claims.

---

# Technology

We are evolving our technology very fast, it feels like yesterday when we use to chat on messenger.com.

 I remember I wanted to come back home early just to come Online and chat with my friends, what happened to our world, it's like suddenly the world we know became very close, everything that happens now in any part of the world does concern us, we share feelings together... I don't feel safe anymore, or private, we are sharing our secrets with people without feeling it.

The worse part about this technology is, if we don't become a part of it, then we are not a part of the society,  it's like you have to have an email or Instagram or whatever you want to call it, and if you don't, you're just a desperate man who has nothing to do in his life.

Just a few days ago, I got a phone call from the ministry of electrical, telling me that they will stop sending my bills to my house address, and that they have a new electronic billing system, they asked me for my email, their new system is sending bills on emails, it's their new way of communication I guess.

It gets very creepy when I think about it, that all of us are nothing but a Data in the system, just numbers on the keyboards.

The question that I always ask myself, what comes after technology?

We won't last too long, the day will come soon, the day when we will go back to the Stone Age, we haven't been treating the earth in a good way, today we are here, tomorrow we will die, and our nation will be a history.

All of these new architecture and buildings around us, someday in the future, will be looked at like the way we look at the pyramids.



The new people of the coming nations will ask the same questions that we ask now: how could they build this technology? How did they survive?



But this is the rule of earth, it has been in this way since the beginning of time, every nation believes that they are the greatest nation of all nations. And they believed that they are the last nation on earth…



Did you really think that this technology surrounding you will last forever?

100 years ago we never had computers, all of these technology we have was built within a century, compare it to the age of the earth, then you have the probability that our technology might not last too long, our technology is like our life time on this earth, its only a matter of a time, till it dies.


Our technology was built on our natural resources, which means eventually it will come to an end, everything on this life is mortal.

# The Existence of Aliens

The age of the universe is estimated to be 13.8 billion years old, Scientists know that the universe is expanding, from that big bang in the beginning of the time, ever that since, the universe started moving on and making its way by the well of Allah.

It really annoys me when people are very sure that we are the only living beings in the universe, I believe that out there, somewhere, there is life, a nation.

when I say a life, doesn't mean that I am talking about Aliens, the same aliens we see in the movies, no, it doesn't have to be that way, and even a bacteria can be considered as life out there.

This bacteria can live and grow to be something big, and it can evolve to be a nation, we don't have to call them aliens, but I am sure things like this might be exist out there.

I'm not saying that you have to believe in aliens, I am trying to tell you that if someday we discovered a living being from outside of earth, just don't get panic.

Till this day we are still discovering new type of fish living in the bottom of the ocean, don't get panic if in the future you turn on your TV and found that there is a living being somewhere out there.

It is the ego of human mind, human mind does not want to accept the fact that there are other creature out there, because it might be more intelligent than us.

Although some ancient civilizations believed that there are intelligent outsiders came to visit us time to time, such as the Maya's, they had drawings that might let us think that they were believing in visitors coming from another planet or galaxy.

If we look at this image it looks like the man is in a machine or in a ship and he looks like he is looking through something.

# War of beliefs

We live in a world that we don't want to live with people who are disagreeing with our beliefs.

Profit Noah was building the Ark in the middle of nowhere, and his people were making fun of him only because he had a believe that it would rain while his people didn't believe that...

When you live in a society that doesn't want you to think as an individual, then you are in a jail, only this jail doesn't have bars to feel it with your hands, you feel it with your mind only.

When your society wants you to have a limited question and thoughts, then maybe you should stand up for yourself and talk, maybe you are the person who can make a difference between them.

I am not saying to be different than them, I am just saying; try to be unique between them.

# Evolution by natural selection: the theory of Charles Darwin:

I would like to talk about someone who really made a difference in our life, someone that stood up for his beliefs while almost most of the people at his time were against him.

He stood up for his beliefs and talked to them, that man is Charles Darwin, who is he, what he believed, and what did he discovered, I am going to mention a short introduction about Charles Darwin, and how did his theory effected the world.

The evolution theory is the simplest idea that anyone has ever had; Darwin opened our eyes on an existed reality that we were not aware of it.

Who is Charles Darwin and how did he discovered his theory of evolution, what it is and why it matter for me??!... This theory is a fact backed with undeniable evidence.

Darwin was born in 1809, his father was a doctor, and wanted his son to follow his career too.  But Darwin was more interested in collecting specimens and shooting. In 1831, he was 22 years old, Darwin got an invitation around the world sailing ship trip for 5 years.

Over the 5 years he collected hundreds of specimens to send it back to his collection at home, he wasn't satisfied with just recording the animals and plants he saw, and so he started to have doubts about the creation of the animals and plants.

He noticed that there are two different type of Ria, but there was a slight difference between them. Did the original group of Ria split into two, and each group got developed in its own way, based on the natural habitat they were living in?

When he returned he was obsessed by what he noticed and wanted to know more… so he asked more. This sailing trip changed him, made his life different, and he saw things that no one else could see.

Darwin discovered that life forms were not fixed, they have changed over time, and they must have evolved.  So he pulled all his collection together to understand why it happened. He got confused that every living thing must be related to every other.  Darwin started to see the history of life like a large family tree.

Some says that the crocodile was some sort of a fish million years ago, somehow it used to come next to the shore to hunt his victims, over thousands of years or maybe millions it felt like it needs to have legs, that will help him to catch his victim faster on the shore.

It evolved to fit the nature of the shore, this fish now is crocodile, now it can live in and out of the water.  Or maybe the crocodile was an animal and now it's an amphibian.

This is just a theory and I could be wrong, I only want to make my point clear.

This is how life evolved, from a single cell, into a complicated creature. You might think it is impossible, but Darwin believed that the cell took for it a massive amount of times to evolve.

45

Everything that we know about life can be explained by the theory of evolution.  It's an undeniable fact that you and I have to live with it, and accept it.

# Evolution

A human is a human, an animal is an animal, and plants are plants, there is no doubt in the creation of Allah, as a Muslim we believe that god first created Adam, and then breathed into Adam.

Everything in this life is a creation of Allah, I believe that the evolution theory does not include in the creation of human, human is a human, from the beginning of time.

But human outfit can be changed and evolve based on the nature he is living in.

For example, why people who are living in the north or south of the earth, are white skinned and blue eyes, this is because their body need to have a blue eyes color cause of lack of sun they have there, their skin is white cause of the coldness.

Imagine if you take a married couple from the Middle East to live there forever, and these married couples are brown skin with a brown eyes.

The first generation will be born there will probably be the same, brown skin and dark eyes, and the second generation also might be the same, maybe even their fourth generation will still have the same, brown skin and eyes

After many generations, somehow the body will start to feel that there is something wrong, and it's the time to change and evolve to fit the nature that they are living in.

Time by time, generation after generation, you will notice a slide different change in their skin color or eyes, or maybe even the color of the hair, still human, only a change in their outfit, to fit the nature they are living in.

The Evolution of every species is the creation of Allah.

Everything and every living being in the universe is created by Allah, and God can create anything in a blink of an eye, but to evolve from a single cell to another, it shows the power of Allah, it shows how Allah is great.

Everything in this life is beautiful, because it's all God's creation, and the universe is just too big and perfect, and a day by day we are discovering a new planet in the space, but unfortunately we choose to believe that we are alone, it's the ego of humanity, not to believe that there might be a living species out of our earth.

Recent discoveries suggested that alien life exists, either in our own solar system or beyond.  Do you know that about 4.5 billion years ago one-fifth of Mars was covered by water, more than 450 feet deep!

If there was an ocean on Mars once, why would it be difficult to believe that there was Life on Mars once?

That means, that there was life on earth the same time as water existed on Mars, because the age of the earth is around 4.5 billion years old.

I believe in the power of God and how great is Allah, and having a beliefs that there is life out of earth will not make me a strange, it will make you a strange that you can see the power of God and still you have chosen to believe that you are alone…

Like I mentioned before, they might not look like how we see in the movies or TV, big heads and green skin, they might be like a bacteria, very small and visible, but after sometime, it will evolve to be something big.

The existence of other species is something with a high probability, only because we don't see them, which does not mean that they don't exist.

If there are other species in the space, what they would look like, what color their blood will be, how they communicate with one and another, how do they eat, what they breathe, do they have the responsibility as much as we do as human being.

Our life is full of responsibilities and if your life does not include one that means you don't have a life, you have to have responsibilities in your life; you have to work hard in your life.

Those species, do they have these responsibilities, do they have bills to pay, how do they survive, do they have ego just like we do, we deny the existence of other creatures, because we want to be the only intelligent creation of God.

The universe is too big, you and I and are nothing in it, there were species on earth since the beginning of the time, only god knows what kind of species existed million of years ago that we can't even find trace for them anymore.

Those intelligence species might visited us from time to time, maybe they use to visit us in the past, then they just got disappointed, they might tried to fix our nation but they have failed, that's why they stopped coming or visiting us anymore, they got disappointed that every time they try to fix our nation, we start another war, war after war, killing the innocent.

I guess they have feelings more than us, because on the earth, they will only feel sorry for you if you belong to their religion or race or a specific group, I guess we have failed the test, we didn't failed now, we have failed since a long time, since the beginning of humanity, maybe that's why they are not coming again.

Or maybe there is another theory about them, maybe they are not coming cause of all of the social media that we have now in our time, it's very easy for anyone to have a photo of them or record them, maybe they don't like to show up on T.V I guess.

But for sure the probability of their existence is high, and if they do, they might be smarter and more intelligence than me and you.

---

# What does Islam think about outsiders?

اللَّهُ الَّذِي خَلَقَ سَبْعَ سَمَاوَاتٍ وَمِنَ الْأَرْضِ مِثْلَهُنَّ يَتَنَزَّلُ الْأَمْرُ بَيْنَهُنَّ لِتَعْلَمُوا أَنَّ اللَّهَ عَلَى كُلِّ شَيْءٍ قَدِيرٌ وَأَنَّ اللَّهَ قَدْ أَحَاطَ بِكُلِّ شَيْءٍ عِلْمًا ﴿١٢﴾

Allah is He Who created seven Firmaments and of the earth a similar number. Through the midst of them (all) descends His Command: that ye may know that Allah has power over all things, and that Allah comprehends, all things in (His) Knowledge.

At−Talaq Verse No:12

$$\text{وَمِنْ ءَايَـٰتِهِۦ خَلْقُ ٱلسَّمَـٰوَٰتِ وَٱلْأَرْضِ وَمَا بَثَّ فِيهِمَا مِن دَآبَّةٍ وَهُوَ عَلَىٰ جَمْعِهِمْ إِذَا يَشَآءُ قَدِيرٌ}$$ 42:29

And among His Signs is the creation of the heavens and the earth, and the living creatures that He has scattered through them: and He has power to gather them together when He wills.

سورة الشورى اية 29

shuraa verse 29

These are the words of Allah in his holy Quran.

And what about IBN Al Abbas, he was one of the Companions of the Prophet (PDUH), that said:

"عَنْ سَعِيدِ بْنِ جُبَيْرٍ، قَالَ: جَاءَ رَجُلٌ إِلَى ابْنِ عَبَّاسٍ، فَسَأَلَهُ عَنْ قَوْلِ اللّهِ: (اللّهُ الَّذِي خَلَقَ سَبْعَ سَمَوَاتٍ وَمِنَ الْأَرْضِ مِثْلَهُنَّ) سورة الطلاق آية 12، مَا هُوَ؟ فَسَكَتَ عَنْهُ ابْنُ عَبَّاسٍ حَتَّى إِذَا وَقَفَ النَّاسُ، قَالَ لَهُ الرَّجُلُ: مَا يَمْنَعُكَ أَنْ تُجِيبَنِي؟، قَالَ: "وَمَا يُؤْمِنُكَ أَنْ لَوْ أَخْبَرْتُكَ أَنْ تَكْفُرَ؟"، قَالَ: فَأَخْبِرْنِي فَأَخْبَرَهُ، قَالَ: "سَمَاءٌ تَحْتَ أَرْضٍ وَأَرْضٌ فَوْقَ سَمَاءٍ مَطْوِيَّاتٌ بَعْضُهَا فَوْقَ بَعْضٍ يَدُورُ الْأَمْرُ بَيْنَهُنَّ كَمَا يَدُورُ بِهَذَا الْكُرْدَنِ الَّذِي عَلَيْهِ الْغَزْلُ" [تَمَّتْهُ إِلَى]

"سبع أرضين في كل أرض نبي كنبيكم ، وآدم كآدم ، ونوح كنوح ، وإبراهيم كإبراهيم ، وعيسى كعيسى" (وقال البيهقي عقبه : إسناد هذا صحيح عن ابن عباس).

A man came to Ibn al Abbas asking him about Sorat al Talaq – verse 12, the same verse I mentioned before, the man asked Ibn al Abbas what is the meaning of the verse, Ibn Al Abbas did not replied and stayed silent, the man said: what is stopping you from answering me?

Ibn Al Abbas replied:  Seven earths in each earth a profit like your profit, Adam like Adam, Noah like Noah, Abraham like Abraham, and Jesus like Jesus.

Now let me ask you, do you believe in Aliens?

If there is seven earths, then the existence of another species there is possible, like I mentioned before, we don't have to imagine them like they are species with big heads and tall fingers, what if… they look just like us, what if they are human like us, they go to work and come back home, get some sleep, what if they have religion and beliefs.

---

# Time travel

Do you believe in time travel?

This question you might hear all around the world, except in my society where this question doesn't even exist, "Do you believe in time travel?"

Why not believe in time travel given all the technology around us that is perhaps on the verge of making it possible?

To believe in a time travel, first you have to understand what is time travel?  Probably you would say: it is to travel to the past or present.

True, but time travel actually is just a concept of a movement, between a certain point in time and another. Unfortunately we cannot achieve it with our technology, because we are surrounded with the limitation of our technology at the moment.

To achieve time travel or to create a time travel, we need a device that runs at the speed of light, and I believe that one day we will be able to create such a device.

Probably you're asking, what the speed of light has to do with time!?

It's very simple, let me put it in this way.

Assume that you are 15 years old and you left the earth on a spaceship that travels at the speed of light, and you stay on the ship for 5 years. Then when you came home, your age would appear to be 20. Meanwhile, you will find that all your friends are 70 years old, probably retired or playing with their grandchildren.

You will discover that the time was moving very slowly for you, while it moved at the usual speed for your friends on earth. While you've spent only 5 years of your life traveling, they have spent their entire life...

now that's a hell of a time travel!

Time travel doesn't have to be by visiting our past or our future, the example that I just showed you is the simplest idea of time travel, and perhaps it is almost a reality.

I am confident that travelling through the time is possible, we only have to improve our technology.

# Relying on Science to Make Sense

Science makes more sense to me than to many other people.

I hate it when I get into a conversation with someone who bases his facts on supernatural accidents or unprovable assertions.

It's literally unethical for you to have a conversation with someone about a specific subject without having proof in your hand to convince that person that you are right about your beliefs.

If you want to have a decent intelligent conversation with someone, always try to put yourself in his shoes. Look at yourself and ask could you actually believe anything that person would tell you without any proof?

Truth doesn't get any more truthful just because you base your facts on what you've heard from others.

The life we live now is different than the past. We can no longer base our truth on something doesn't have a foundation in science.

Anything that science can't find an explanation for, then it's either a miracle of God or a lie.

Having a conversation about something and basing facts on science doesn't only makes more sense to others, but it also shows people how well educated you are, and a great talker. People will just love to learn from you, even if they are disagreeing with you, still they would love to listen.

It's your choice, which type of person you want to be—the guy who thinks he knows everything and bases his information on what he has heard from others, or a guy who talks about things based on what he learned from science and history books.

It's very easy to be the first one, but you'll end up stupid. The second one will take time to learn about life, but you'll end up more successful in your life.

But when you are basing your facts on science, assume the other person is ignorant. That means it is best for you to listen to him, even if you know that his information are wrong, because if you interrupted him, there will be no difference between you and him.

# Between belief and truth

Believing is not only a state of mind but it is also a way to live our lives.

To have a belief, we need Truth, without it the circle of life cannot be completed.

We can't live without having beliefs in our life's, having beliefs is what brought us to create such wonderful technology, so it's very important to have beliefs that keep you moving, learning and growing,

But as I said before, a belief needs its other half, which is Truth.

Because if you belief in something without knowing the truth about it, it will lead you to your destruction.

In between the belief and Truth, there you will discover yourself, and you will find knowledge.

Having a believe with truth in the same time, makes you a wise man, a man who can discuss his opinion about his beliefs with people without getting into a war with the other party.

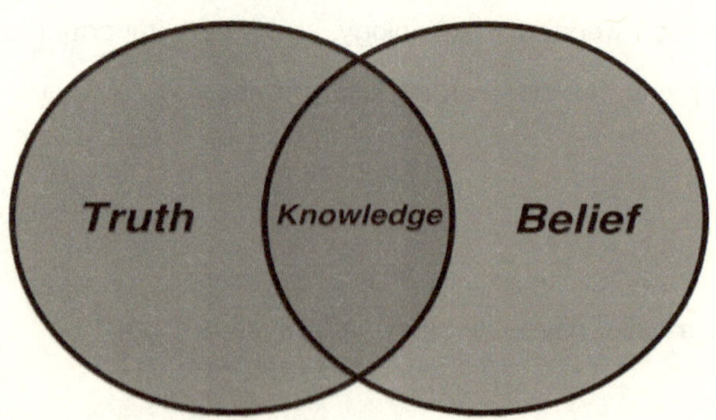

Some people might say, we don't need to have beliefs in our life, we can be enough by knowing the truth, still they are mistaken, they forgot the truth does not means knowledge, and also a beliefs does not means knowledge, we have to combine them together to get the final result, which is knowledge.

Where I come from, some of people believe that children should not play football at night cause they might disturb the people who live under the earth ( Ghosts ), and generation after a generation are having the same beliefs, that we should not play football after 8 pm.

I grew up knowing what we have been told not to play football at night.

Is cause our family wanted to sleep, so people came out with the ghost idea to scare us, there is nothing wrong in playing football at night, it doesn't matter, do not create a lie for your own benefits, someday people will grow, and the lie will grow with them, the lie will turn into a belief.

Give your children believe and truth, and their legacy will be knowledge.

What makes things funny to me, is that people believe in superstitions like that but still they refuse to believe that there are creatures living outside of Earth.

At least I have proofs that life might exists in the space, what is your prove that ghosts are not allowing you to play football?

# Between Fear And The Reality

We all hear sometimes strange voices when we are alone, or when we are a little bit afraid, straight the way we think it's the sound of ghosts, but have you ever asked yourself... Why?!

Why would you think that this strange voice or noise was made by the ghosts? I can tell you why, because since when we were kids, our parents put in our minds the idea of ghosts, that they are able to haunt our bodies, that's why when we grow up, the idea grows up with us.

The truth is, they only haunted our mind...

That's why when you are alone in the dark, you keep on thinking about the ghost stories, and the possibility that they can come for you in any second, when you keep on thinking in things like that, will bring negative thoughts to yourself.

Negative thoughts can create strange voices in your mind, sometimes it can create fake images too, keep in mind that next time the sound that you will hear when you are alone, is not the sound of ghost, maybe you just forgot to close door to the patio.

"You only hear what you want to hear."

I remember when I was a kid I use to get thirsty in the middle of night, and I wanted to go to the kitchen to get some water, but it was too dark and I was a little boy who was afraid from the dark.

I thought that there is a beast waiting for me in the kitchen to take my life, time after time when I got bigger and became a grown up man, I started going to the kitchen in the middle of night, and there was nothing there, just a kitchen with lights OFF, and all what I was thinking of was those fake stories about the ghosts.

Now I'm 26 years old and I'm so mad how could I hold my thirst for too long just because I've heard a ridiculous ghost story that wasn't even true.

I was simply believing something based on other people's lies, but when I grew up I saw that the kitchen is still the same.

There is nothing wrong about being alone in the dark, the difference about the light and dark is, that in the light you are able to see everything clear, and in the dark, you are not able to see clear, or maybe you won't even be able to see at all.

We all have fear living inside us, we can't deny that, but sometimes we don't use our fear in a correct way, for example, we can use fear to be afraid from our mistakes, and that one day God will judge us, but still we choose to fear from unreal stories about ghosts.

The moment you communicate with fear, is the moment that you overcome it.

If you are so afraid of doing it, then maybe you should try it.  But still we need fear in our life's, without it we might get lost and do wrong things, things that can destroy the people around us, everything we do, it effect the people around us, it won't only effect you.

Fear can bring nothing to your life but negative ideas that leads to failure and your destruction.

A friend of mine once told me his story about going into a haunted house, he kept on repeating his story for years, and he insisted that he saw a ghost there.

According to what my friend told me about his story, I believe that when he decided to go inside the house he was scared and panic, that's why the negative thoughts affected his brain. He thought that he saw something strange in that house.

It might be true, but I doubt it.  I would guess that it was just a part of his imagination that he saw a ghost.  In another words, we can say that my friend had negative thoughts more than the positive thoughts.

When you think negative, bad things starts happen to you, but when you think positive, positive things start happen to you. My friend went into that house with a lot of negative ideas and thoughts, that's why his mind automatically created a ghost image.

Some people say they believe that two spirits are possible to join or to be in one body, but what I believe, is that two souls can never combine and join to be in one human body.

God created us perfectly, having beliefs that a ghost or an evil spirit can haunt someone's body is totally unbelievable for me, God created us perfectly.

If ghosts are possible to see, then why it's not possible for me to see them? Are those people better than me?

Or maybe those people who are pretending that they have seen one, they just made it up!

Maybe it's just their fake story to get some attention from others?  I am not saying that they are liars, I'm just saying what if they have some mental problem…I do believe in ghosts, and I do believe that they exist, but I don't believe that it's possible to see them.

# Possessed by evil

Long time ago, when a family member use to get some kind of disease, his family used to think that he is possessed by evil, they use to hit him badly to suffer the evil spirit.

They were wrong, this person was just having Epilepsy, and there was no evil spirit inside of him, they were biting their poor family member only because of what they were believing.

It's fine to believe in anything you want but make sure that your beliefs will not harm the people around you.

We are the one who are haunted by evil, we create our own hell... we shall blame ourselves for our faults.


We created every negative thing on this life, we are haunted by our own ignorance, but unfortunately human beings avoid these mistakes and seek out other's mistakes.

---

□

# Unleashing yourself

In life, people are always afraid of opening a door that no one has opened before. Most of us doesn't take risks to open this door, because of the stories we have been told or because a specific belief.

But we have to take risks, life is always about taking risks. If you want to be something in your life, then you have to take steps forward. Each step opens a new door in front of you, a door that you don't know what is hiding behind of it.

Forget about any stories you have been hearing since you were a kid! "Don't go inside, the house is haunted."  Instead, I say that you should not judge the house from the outside, go inside and take a look...then you will be able to judge it.

It's very sad that we always choose to believe what we heard from other people without proofs.

Life is about taking risks, and adventure, and if you want to have a limitation of thinking like some people, then believe any story you hear from anyone, but before I leave you alone with your believes, let me tell you something, we are living in a modern century, we are basing our facts on science.

Make your choice, open the door and find out what is inside, seek knowledge, or else stay on the same level as you are on now, and if you want to live your life in this way, then you will never move on.

Others have opened many doors in their life, and that's how now they are successful.

You only need to use one more percent from your brain... to change your entire life.

You're the owner of your life, play it right, don't let obstacles hold you back, 20 years from now you will regret the things that you thought that you are not able of doing it but actually you could if you had the courage.  Don't let your specific belief about something hold you from moving forward.

---

# Is Islam against Education or knowledge?

This topic is for those who believe that Islam holds us back from moving forward.

Knowledge is the key, knowledge is our salvation.

Our religion started with Eqra, which means READ in English.  Islam is knowledge, Islam is the Key, for your salvation.

In Bader, the Prophet Mohamed commanded each prisoner to teach 10 Muslims how to read and write in advance to set them free.

## Is Islam against education?

The answer to the question is defiantly NO, that's why have you ever wondered why many stars have Arabic names? Or why there is a science called Al Jabr?

Unfortunately people now believe that if you want to evolve to get more knowledge or to be a scientist, then you have to let go of religion, because religion limits your thinking, as if religion does not allow you to develop your knowledge. However, this is not true. Islam and most other religions encourage people to read, learn and study. No matter how far it gets, Islam still encouraging us to learn and ask more to learn.

"اطلبوا العلم ولو في الصين"

Islam doesn't want you to believe that you were only born, it wants you to learn how you were born, " the stages " and the stages is written in the holy Quran, since 1400 years ago.

سورة المؤمنون – Al-Mumenoon: Verse 14

ثُمَّ خَلَقْنَا النُّطْفَةَ عَلَقَةً فَخَلَقْنَا الْعَلَقَةَ مُضْغَةً فَخَلَقْنَا الْمُضْغَةَ عِظَامًا فَكَسَوْنَا الْعِظَامَ لَحْمًا ثُمَّ أَنْشَأْنَاهُ خَلْقًا آخَرَ فَتَبَارَكَ اللَّهُ أَحْسَنُ الْخَالِقِينَ.

Then We made the sperm into a clot of congealed blood; then of that clot We made a (foetus) lump; then we made out of that lump bones and clothed the bones with flesh; then we developed out of it another creature. So blessed be Allah, the best to create!

Names of some famous scientist from the golden time of Islam:

1-	Al Biruni, Died in year 1048, he was one of the firsts to prove that the earth is rounded, using the Holy Quran.

2-	Al Farabi, died in 950.

3-	Jabr Ibn Hayyan died in 815, and he was the first to talk about the chemical reactions.

4-	Al Hassan Ibn Alhytham died in 1040.

5-	Al Khwarizmi, died in 850, Al Jabr, and until this day we are studying it.

To follow the path of Islam, it requires all Muslims to have knowledge about two things

1-    Prayer time, for example, we pray five times a day, but there is a certain time to pray, a specific timing, based on the movement of the sun.

2-    The direction of Qabah ( the house of God ), all Muslims pray in one direction, which is Qabah, and that's how many Muslims still if they are in the middle of the desert can easily find the direction of Qabah.

What we learn is that all that time people were wrong about the idea that a religion can hold someone back from learning or developing his knowledge, Islam encourage us to learn unlimitedly.

# The Man of Salvation

The earth estimated age is 4.5 billion years, and still much of human civilization believes that somebody will come to save the day.

Why are we living in a pathetic world to believe that someone will come and bring salvation with him, why does it have to be in that way?

Can't we be responsible for what is going on in this world, can't we start the change through ourselves and our own actions first?

In every time of this life, in every civilization, it was believed that there would arrive a man who will save the day. He will take the revenge on those who killed the innocent, raped women, and stole from the poor.

People waited and waited, and that man never showed up in their time, because his time didn't came yet.

Someone will come someday but not yet …

But we can't just get killed and remain silent and wait for someone to come to rescue us!

The worst part is that our civilization tends to believe that we are the last, and there will be no more after us. This implies that the end of the Earth or the judgment day as people call it, will be in our time.

The world will not end in our time, the world will continue like always, this is the nature of the earth, it heals itself after man's destruction.

Just as always throughout history, humans cause damage to their world, such as war, killing, crimes and environmental destruction. Yet our world continues and another civilization will come. They will see our work, our cities and all that we have constructed. They will probably look at us the same way we look at the old civilization of Egypt.

There is no doubt of the end of the time, there is no doubt of the judgment day, but it's the ego of mankind that doesn't want another civilization after him, just like denying aliens, mankind doesn't want to believe in other species like the UFO's only because they might be more intelligent than us.

After our civilization, Blood and war will be our legacy, and then the new civilization will come, starting the earth from zero, no electricity, no oil, and no weapons of man destruction.

Their world will be bigger than ours, thousands miles of travelling on the ground.  No more boring times, no more family time watching superman movie on the T.V.  Just a pen and a paper teaching their children's how to survive.

No Google, no social media will exist, just a dusty library in the house of their old man that lives next to the river, that river that once was a bridge there, connecting two beautiful countries together, before the lie comes in between them, and destroys them, with the big lie, they lived to fight for that lie, and died for nothing.

This is how we created hell in our time, we created lies and sow discord among the people, for our benefits.

Each one of us believes that he is right and correct about what he believes in, with all respect to your religion, but your religion should have nothing to do with my death, I should die in a car accident or any other normal type of death, than dying cause a suicide bomber or a war victim under the name of God.

There is no chance and no right for you to kill me only because we have different beliefs, we all belong to earth at the end, we all have one father and one mother, we are all fighting to survive this life in a peaceful way, without harming anyone.

We should not belong to a specific race or group, we should all be considered as human beings.

They are fighting under the name of money and not God, God in all of his holy books, told us to live in peace and love.  No matter what you believe in, God wants you to live in peace.  Our prophet Mohammed peace be upon him, was the master of peace, a teacher of forgiveness.

# Believing in Fear vs. Taking Control of your life

I have been trying my whole life to challenge the fears that I have, and I have discovered that losing what I have built in the last couples of years is my biggest fear of all.  Sometimes I wish darkness was my biggest fears, because it will be an unreal fear, or a temporary fear, which means it's only a matter of hours till the sun comes out, and after that the shadows will disappear, light will come back again, and fear will be defeated.

I am not afraid of evil… dark spirits as people say, I'm not afraid of being cursed by some ugly witch… I don't believe in that stuff anymore.

I have sat in the dark for hours and I have seen nothing, heard nothing.

I have decided to believe in reality fears, such as losing people or the fear of failure these are beliefs that I need in my life.

The fear of losing what you have built is a real fear and a real belief. Now I have my own empire and I have to protect its property, and I must make sure not to get destroyed.

Everything might change within one night, and only because you have built an empire doesn't make you immortal.

As I mentioned, the objective of this book is not to show you that there is many kinds of beliefs. Rather the objective is to believe and move on, without harming anyone.

I have meet a lot of ignorant people in my life, and most of them I am still dealing with almost every day. They make me feel sick, mad, and insane, but I go on with life. If life wants to play a game with me, then sure I am ready to hold the joystick of life. There is no honor in an easy life.

I believe that the most important thing in life is taking control of your life. You are responsible for it. Don't blame someone else for the life you've got. If you are poor, get up and do something.

I know lots of poor people in my life, and they are very good to me, but they have one problem-- they are waiting for a job to find them, they are mistaken, they need to believe that they can find a job and work harder to find it.

This is the reason why only few people change for the better in this world. Too many people are waiters, meaning that they wait for change! You'll never find a rich successful man saying: I've waited to collect money as much as I could just to get rich!

You'll never find someone say that, because part of success is to move on and try make a change, and the main function of success is Belief.

I can't imagine why people only rely on one job, or one income in their life, don't you fear losing your career!

Like I mentioned, people choose to believe an unreal beliefs, like that their job is immortal.

If someday, God forbidden, I lost my job, I'd never blame it on the government, or I'd never expect them to find me a job immediately.

I would only blame myself not to have another income to relay on just in case if I lose plan A.

# The Earth

This is earth,

A planet that lives on it every thief and oppressed, every rich and poor, every believer and godless, every weak and strong.  From the beginning human use to have different beliefs between their self's, which led them to war.

War existed from the day Adam came on earth, between Evil and Good…

And here we are today, in a time of technology, the century of evolution, which our nations are educationally close to one and another, but far from one and another in their opinions.

Between us are nations that set their life's in peace and development, and also there are nations that set their life's to reveal other people's mistakes or disadvantages, basically they are doing nothing to improve our world.

There are 7 continents in our planet, and we have about 231 counties, and we also have many religions and beliefs, many races and languages.  We are more different than we are similar… I guess we are only similar in shape.

# Conclusion

We can't live an individual life, no matter how much independent we can be.  Mankind does not exist until he is gathered with his own species. Mankind needs cooperation, but cooperation needs negotiation first.

Our unity is our strength, and strength leads us to fight, fights leads to peace or destruction between nations, and that's what makes us human being. No matter what we agree on to believe, still some day we will be divided, nothing lasts forever, even beliefs will come to an end eventually.

Being individual and independent in your beliefs is a valuable thing, but not after some time, because you will feel lonely someday and that you are different, and unfortunately you will look for an escape, just to run away from your beliefs, because your beliefs were the reason for your loneliness.

Sad but true…

If you are looking for a belief to hold on to it forever, then try to be united with people who have the same beliefs as you do.

We are human, we fight, we love, and we belief, but even if your beliefs does not makes sense to us, still we have to respect you and your beliefs, as long your beliefs do not harm anyone, we don't have any reason to stand in your face.

This is inside your mind, explaining for yourself what you've done to your world...

This is inside their mind, explaining to themselves what they have done to their world...

This is inside our mind, explaining to ourselves what we have done to our world...

This is ....

Inside A Human Mind...

This is inside your mind, explaining for yourself what you've done to your world...

The End

رقم الناشر الدولي :

978-99958-93-93-4

رقم الايداع بأداره المكتبات العامة:

61/د.ع/2017

9 789999 589394